Draw Robot's meal.
Robot eats lots.

My robot had 

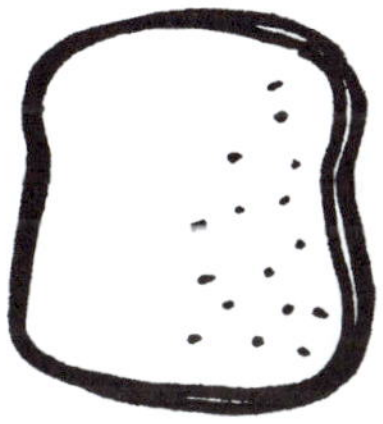 pieces of bread

and

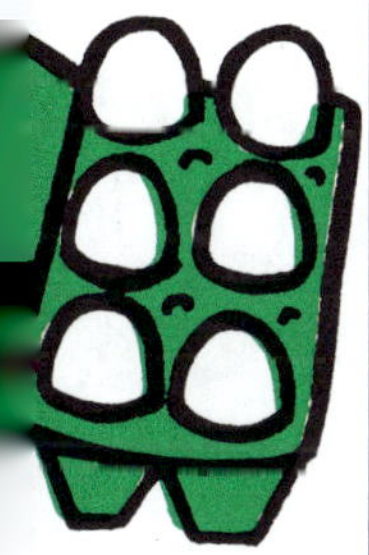

I can count to

Note for families: This activity is designed to help your child to think about numbers. Even if children cannot as yet do addition and subtraction very well, they enjoy thinking about large numbers. The border contains things for your child to count with you. If they can do this easily you could help them to see how far they can count. Let them have fun with this! They might want to try to count to a million!

to school:

Comments:

............................and ....did this together.

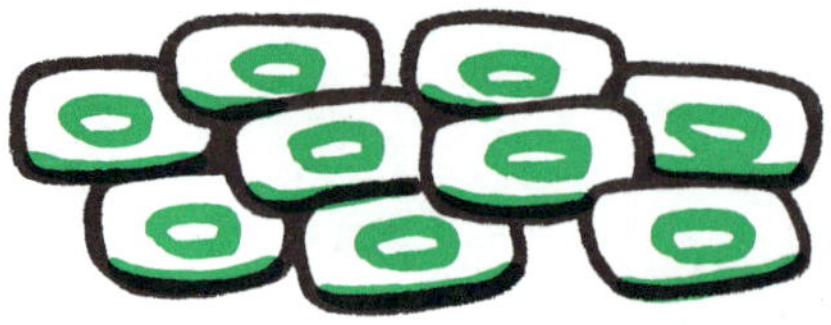

Note for families: To play this game you need a 1–6 dice and some kind of tube (a kitchen roll middle or rolled up newspaper). You can try holding the tube at different angles to aim at a number.

Game 1. Take it in turns to roll the dice down the tube. If the number on the dice is the same as the number on the space it lands on, you score that number. So, if you roll a 3 and it lands in the space with 3 in it, you score 3. Write down your scores. The first to get a score of 15 or more is the winner.

Game 2. Race to score all the numbers, 1 to 6.

Game 3. In this harder game you don't need to match the dice to the picture. You add the score on the dice to the number in the space it lands on. So if you roll a 4, and it lands on the 6, you add 4 and 6 to get your score of 10. Race to get 100.

The border is for counting and writing in the numbers, eg put a 10 on the space by the 10 books. You could add up everything in the border.

Comments:

Date to schoo

............................and ....did this together.

Draw some needlework things you saw in the shop or at home. You could put some price labels on if you want.

Note for families: To help the children with measuring we will be looking at things sold at a sewing counter. You could help your child by

- looking at buttons, needles, etc in shops,
- letting your child play with the button box,
- measuring lengths of wool, ribbon or string with a tape measure,
- sending some items to school (eg a length of left over ribbon, buttons, balls of wool, empty cotton reels),
- talking about the items in the border and what they are for.

to school:	Comments: ....and ....did this together.	1.4

Ask to see the coins in someone's purse.
Draw round the coins in the right space.

	silver	not silver
not round		
round		

Note for families: We are learning about money at school and this sorting diagram might help your child to recognise and name coins. Each coin belongs in one of the sections in the diagram, eg the 50p coin would go in the top left since it is silver but not round.

Border activity: How many 1p coins are there? How many 2p? 5p? 10p? 20p? 50p? £1? Some children have a clear grasp of money amounts and with help they might be able to add some of the coins in the border. Children often need considerable help with totalling amounts of money especially if they are not used to handling money.

Comments:

Date to schoo

............................and ....did this together.

altogether take away plus add

×

difference between

=

6 − 1 =

8 − 3 =

2 + 4 =

−

4 + ☐ = 8

Mystery sums

You need: a pack of cards with the picture cards taken out
a 1–6 dice.

Game 1. Shuffle the cards and put them in a pile face down. Take turns to take a card and keep it hidden from your partner. Roll the dice, add the total, and tell your partner. They must now work out the number on your card.

Game 2. This is played like game 1, but this time you work out the difference between your two numbers, for example if your card is 2 and you roll a 5, the difference is 5 – 2 = 3.

Note for families: These games will help your child to add and subtract numbers. They will be helped if you talk about numbers with them, and use some of the words in the border. If you don't have cards, you can make cards 1 to 10 from a cereal packet, and you could make a second set 1 to 6 if you don't have a dice.

to school:

Comments:

..........................and ....did this together.

Dragon twenties

You need: a pack of cards with the picture cards taken out
a dice
paper and pencil to keep score
2 counters

Shuffle the cards and put them in a pile face down.

Take turns to turn over the top card and roll the dice.

Choose just one of the numbers to add to your score.

When you score exactly 20 shout dragon!

Border dragon race

Put your counters on start.

Take turns to take the top card and roll the dice.

Take one number from the other and move your counter that many dragons round the border. Race back to the start.

Note for families: These activities will help your child with addition and subtraction. If you don't have cards, you can make cards 1 to 10 from a cereal packet, and you could make a second set 1 to 6 if you don't have a dice.

Comments:

............................and ....did this together.

Date to scho

Draw some shapes you found that fit together.

Note for families: Some examples of shapes that fit together are shown on this page. You and your child might might be able to find more. This will give you a chance to talk about shapes. Your child might talk about tessellating shapes and that means shapes that fit together like tiles on a wall. You might want to help your child with their drawings or find some pictures of patterns (eg in wallpaper books and on gift wrapping paper.) These can be taken to school so your child can show the other children.

e to school:	Comments: ....and ....did this together.

11 12 13 14 15 16 17 18 19 20 21

10 22

9 23

8 24

7 25

6 26

5 27

4 28

3 29

2 30

1 31

Start 32

Find something you can put into sets of two. For example  cutlery  gloves or socks;

Ask someone to help you to count in twos: 2, 4, 6, 8, . . .

We counted up to _____ in even numbers.

Even number dot to dot

Watch out! You must join the even numbers!

Border game for 2 players

You need: any dice or a pack of cards with the pictures taken out, 2 counters, lots of buttons (or raisins).

If your child is not yet confident with odd and even numbers, you could ask them to put a coloured ring around the even numbers before you play.

Take turns to throw the dice and move your counter that number of spaces. If you land on an even number you win a button. The winner is the player with most buttons when you both get to 32.

Odd
1, 3, 5,
7, ...

Even
2, 4, 6,
8, ...

Note for families: We are talking about twos in school and it will help your child if you can count with them in twos. You might be able to count to over a hundred together.

Comments:

Date to school

............................and ....did this together.

Design a pattern on this T-shirt.

Make these match the T-shirt

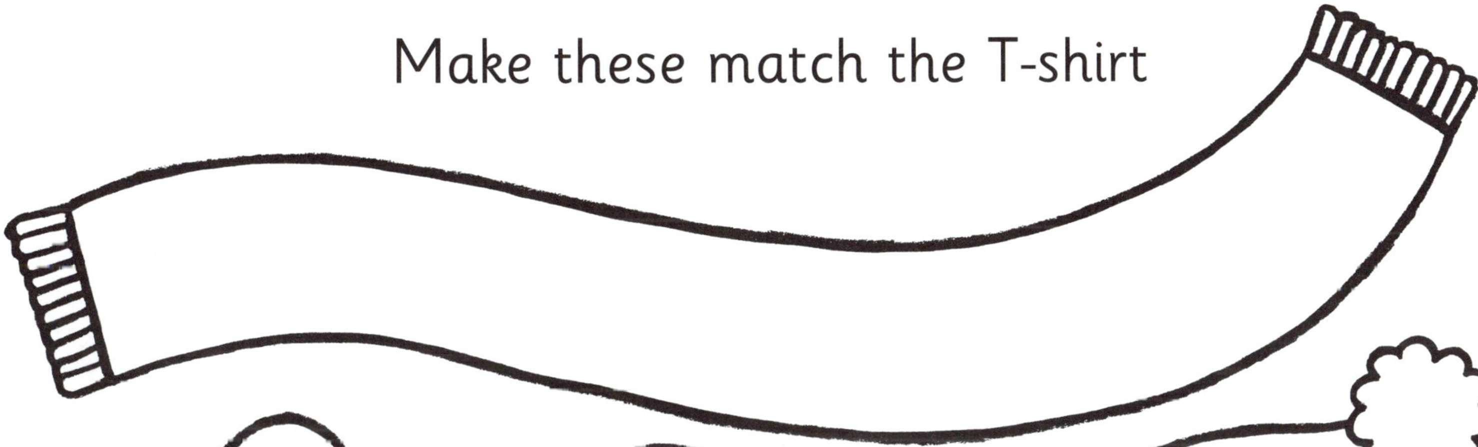

Note for families:

Please help your child to find patterns on clothes such as patterns on jumpers and socks. It is particularly helpful to look at repeated patterns such as green circle, red square, green circle, red square, etc.

Help your child to choose just two or three colours to make their patterns on the T-shirt. They may find it helps to try out a pattern on scrap paper first.

to school:	Comments: ....and ....did this together.	3.2

12:00 day early morning 10:00

afternoon 1:00

late

3:00

a.m.

9:00

midnight

7:00

6:00

noon

night

8:00

p.m

day:	day:
:	:
:	:
:	:

"I got up"

Note for families: This page will help your child to recognise hour times on both analogue and digital clocks. Many children find it easier to tell the time on digital clocks, but it helps if you have a clear analogue clock and ask your child to read it at meal times or story time.

Your child can record the time they wake up and two other events for two days. They will probably need help to write these times on the clocks.

Border activity. There are eight times shown on both analogue and digital clocks. Ask your child if they can find two 6 o'clocks etc. There are also some words describing time and it would be helpful to talk about these with your child.

Comments:

............................and ....did this together.

Date to school

Jack's game

$3 + 1 + 2 = 6$

$2 + 2 + 2 + 1 + 3 = 10$

makes

add

4 + 3 2 =

My 3 numbers are

We made these sums.

Note for families: We have been doing some maths from the story of Jack and the beanstalk at school. This activity will help your child become more familiar with numbers and have fun doing maths.

Jack's game

You need dried beans, or other objects to help with the addition. Choose 3 different numbers. Try to make as many different 'sums' as you can, using your three numbers as often as you want. Some children will need plenty of support to do this. Most children can do maths in their head and with counters before they have the ability to write it down so you might want to write for your child.

Your child might want to continue the game with three different numbers, or use a calculator to try to find what happens with large numbers.

Count the number of items in the border.

to school:

Comments:

..........................and ....did this together.

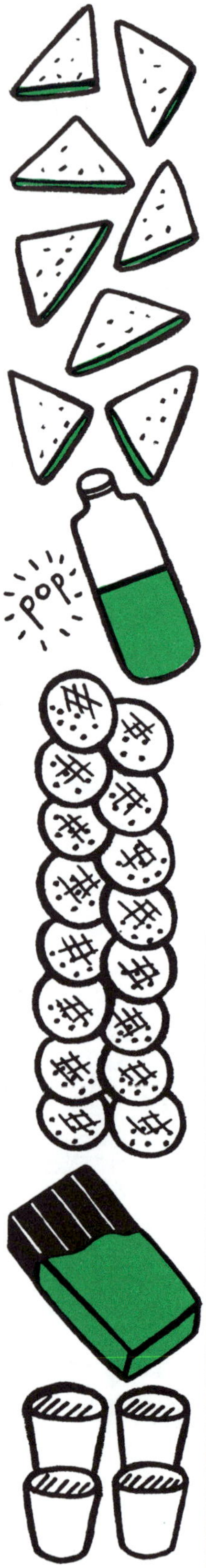

Jack's party 

Jack invited 3 friends, wizard 

princess  and slippery snake 

Can you share out the pizzas?

How much each?

Can you share out the things in the border?

Note for families: we have been doing some work on halves and quarters in school and you can help your child by talking with them about these. For example, you could:

- halve an apple, potato or tomato
- pour out a quarter of a glass of squash
- look for other halves and quarters at home or in shops (half a kilo of apples etc).

It is some time before children use symbols such as $\frac{1}{2}$ or $\frac{1}{4}$ but we are gradually introducing them. Your child might well talk about 'the bigger half' and you need to remind them that a half is exactly the same size as the other half.

Use the border to talk with your child about sharing out the things for Jack's party. Most things will share out equally but there is one balloon left over.

Comments:

Date to scho

...........................and ....did this together.

Ask someone to help you to count to a hundred.

Ask someone to help you to find more large numbers.
We found these large numbers.

My largest number was ______

Giant numbers game

You need: a pack of cards with all the pictures taken out (or two dice)
some dried beans or Lego to count with

Take two cards each, turn them over and try to make the largest number you can. For example, a 4 and a 5 would make 54.
The player that makes the largest number wins a point (a bean).
The winner is the one with the highest score at the end of ten goes.

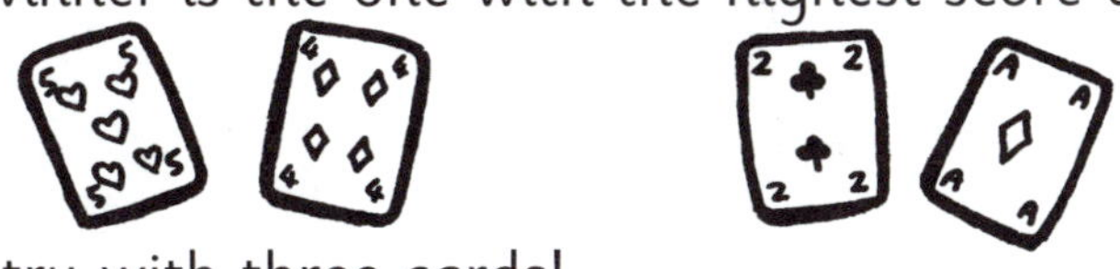

54 beats 21.

Now try with three cards!

Note for families: We have been working on large numbers at school and these activities will help your child to count to a hundred and to learn something about numbers beyond a hundred. If your child finds it hard to make large numbers with playing cards, you could make simple number cards from a cereal packet. You can talk about the large numbers in the border.

to school:

Comments:

..........................and ....did this together.

Published by the Press Syndicate of the University of Cambridge
The Pitt Building, Trumpington Street, Cambridge CB2 1RP
40 West 20th Street, New York, NY10011–4211, U.S.A.
10 Stamford Road, Oakleigh, Melbourne 3166, Australia

Sue Atkinson Sharon Harrison
Lynne McClure

Illustrated by Sonia Canals

First published 1995
Printed in Great Britain by Scotprint Ltd, Musselbrugh

CAMBRIDGE
UNIVERSITY PRESS

ISBN 0-521-47592-9